How to use this book

A sample page

INSTRUCTION
What your child needs to do for the activity.

TITLE
The page title describes the skill your child will learn in these pages.

FUN ILLUSTRATIONS
Specially drawn illustrations which are fun, interesting and drawn at the right pedagogical level for your child.

COLOURFUL BORDERS
The page borders make each page as attractive as possible to stimulate your child.

EXAMPLE
The first one is done for you so you can show your child exactly what to do.

LOTS OF PRACTICE
Two pages where your child can practise and repeat the same skill to master it.

STICKERS
Place a sticker on each page as your child finishes.

WHO'S HIDING?
In each book, a little creature appears in the border of every double page so your child can have fun trying to find it.

EXTRA ACTIVITIES
Extra activities you might want to do with your child to further reinforce the skill or simply make it more enjoyable.

Step-by-step learning

STEP ONE **Read** out the title of the activity page to your child.

STEP TWO **Explain** the skill and show your child the example already done. **Make sure** they understand what to do. Your child will then have at least two pages to practise that same skill.

STEP THREE **Help** your child put a **sticker** on the bottom of each page as they complete it.

Remember to be patient, encouraging and positive with your child, even when minor mistakes are made!

How to hold a pencil

It is important that you help your child hold his or her crayon or pencil in the correct way as shown here to ensure your child develops the right technique early on.

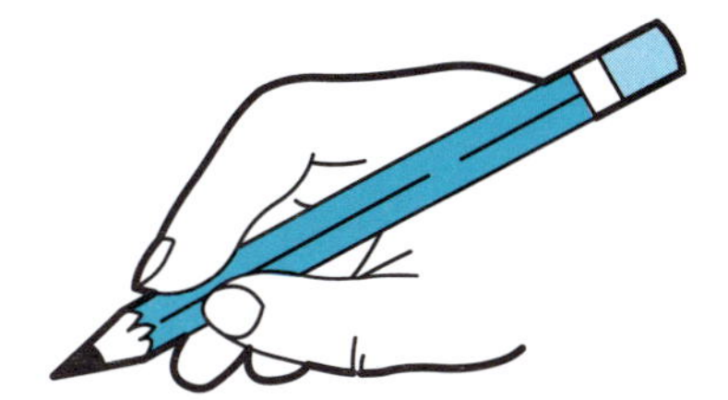

Finding the shorter one

Look carefully at each pair. Which picture has something shorter? Circle the shorter one.

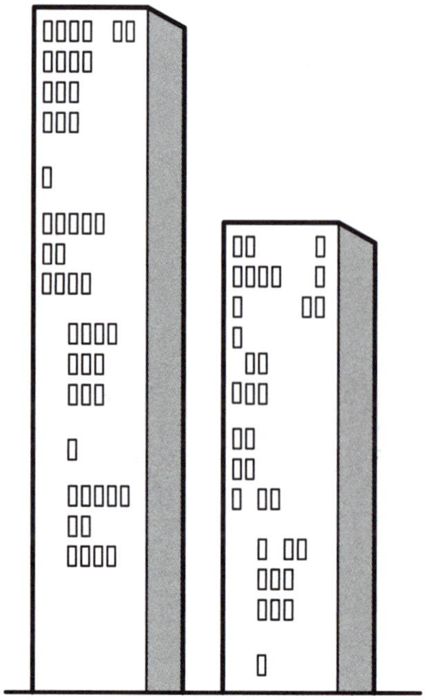

1. Cut a piece of string as tall as your child. Ask them to find 5 things which are shorter than this piece of string.

Place a sticker here.

2. Select 3 objects of different sizes. Ask your child to place them in order from shortest to longest.

Place a sticker here.

Drawing one shorter

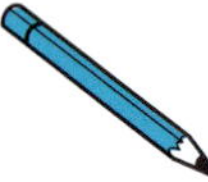

Look at each object. Draw a shorter one next to it.

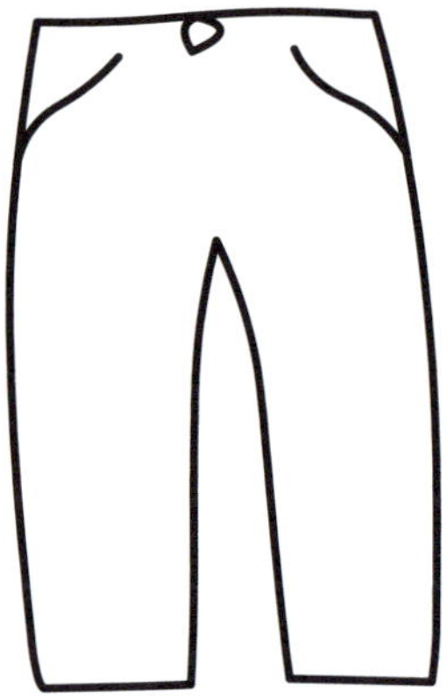

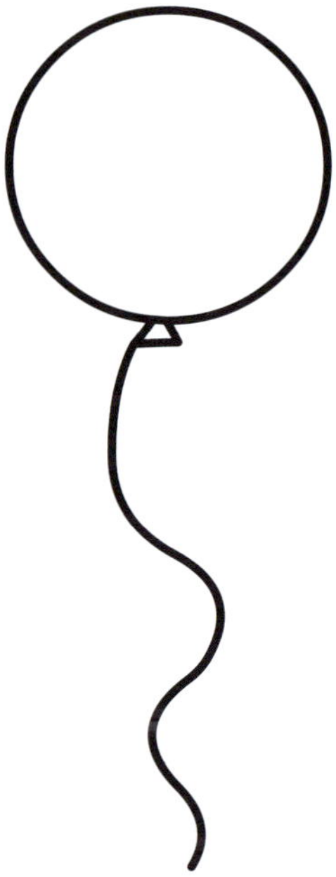

1. Ask your child which is longer: their arm or their leg? Can they guess which one? Help them measure by cutting pieces of string the length of their arm and leg.

Place a sticker here.

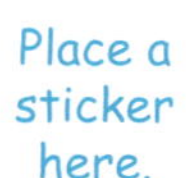

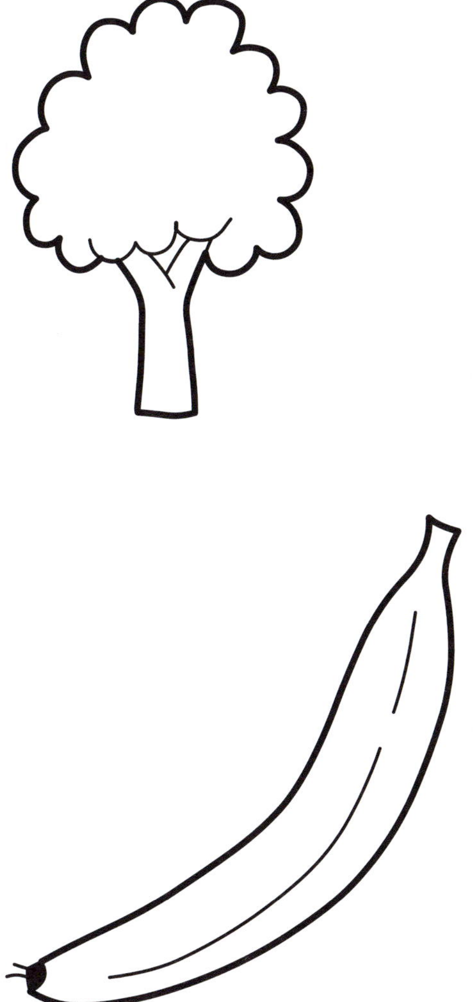

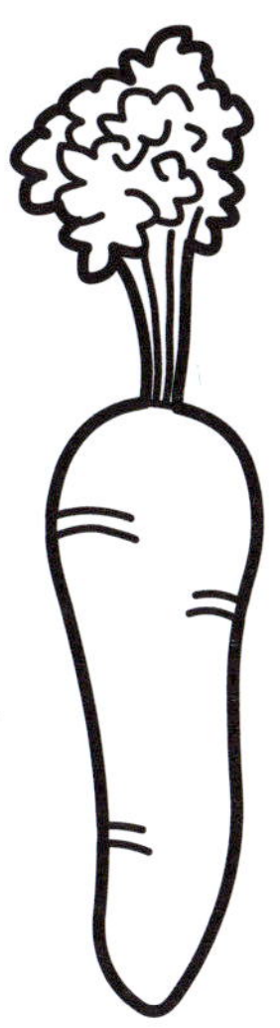

2. Use a streamer or string to measure your child and yourself. Who is shorter? Who is taller?

Place a sticker here.

Finding the smaller area

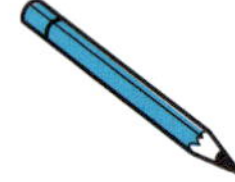

Look at each pair of pictures. Which one covers a smaller area? Circle the smaller area.

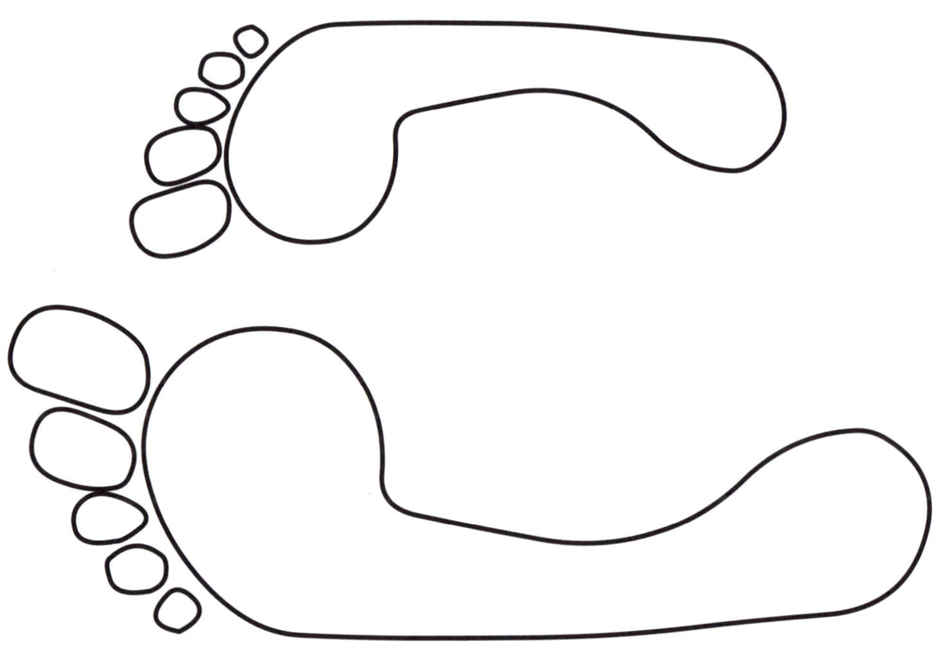

1. Ask your child to colour in the pictures which cover a smaller area on these two pages.

Place a sticker here.

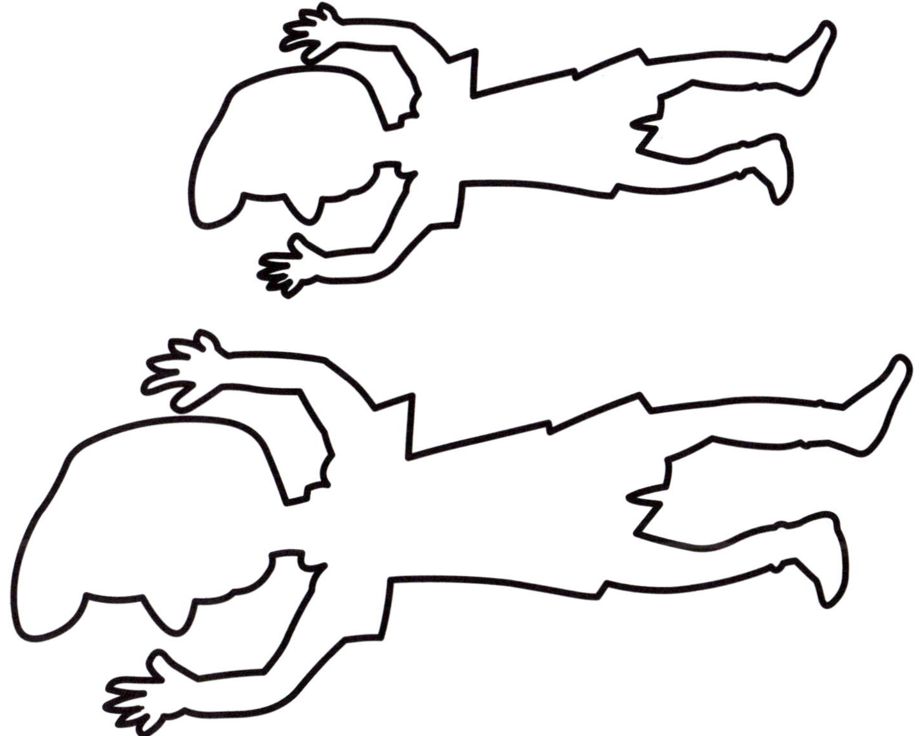

2. Help your child trace around their hand or their foot on paper. Then ask them to find flat things which have a smaller area than their hand or foot.

Place a sticker here.

Drawing a smaller area

Look carefully at each flat object. Draw one next to it which covers a smaller area.

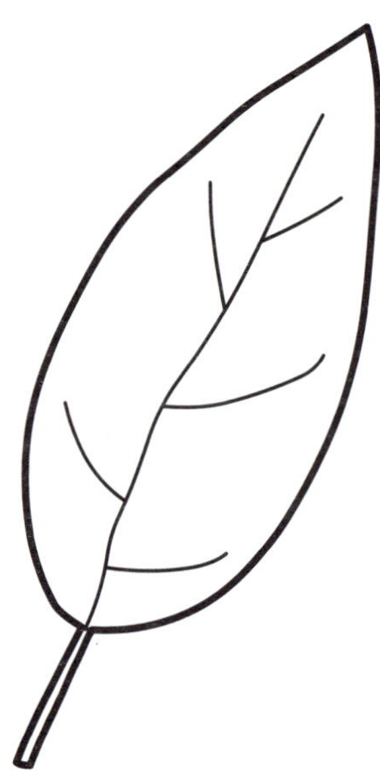

1. Ask your child to find other flat things in their room and look at their size. Which ones cover the smallest area?

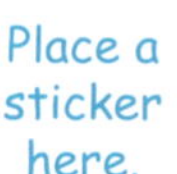

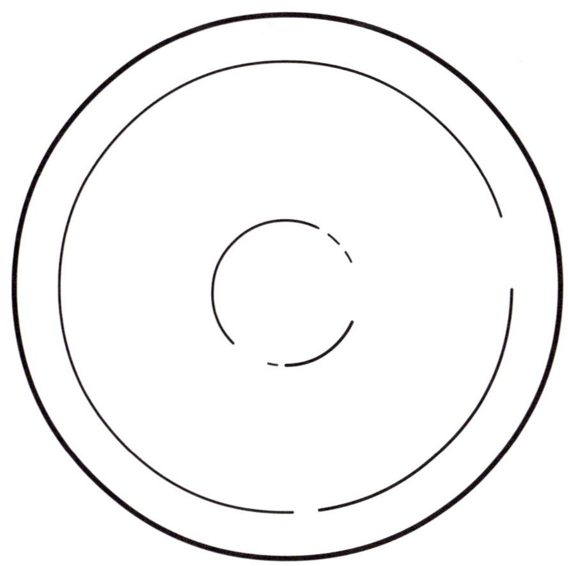

2. Select a group of flat objects (eg. books, cards, plates, etc) and ask your child which ones have the largest and the smallest areas.

Place a sticker here.

Finding one smaller

Some things take up less space than others. Look carefully at each pair. Which one is smaller? Circle the smaller object.

1. Try to find pairs of objects around you at home and ask your child to point out the smallest one in each pair.

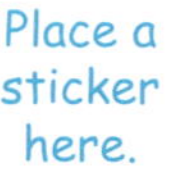

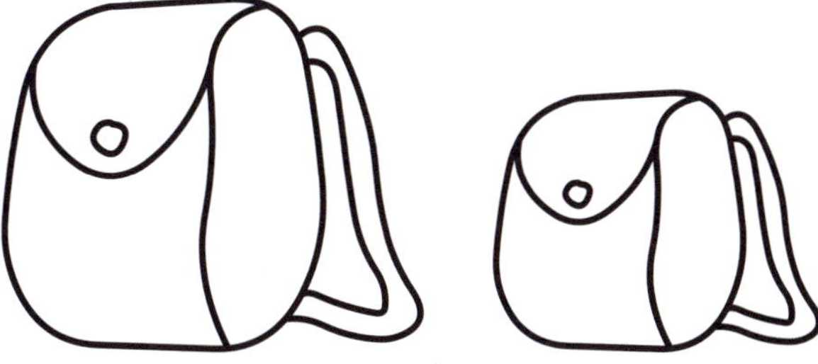

2. Find some boxes of different sizes, and ask your child to put the smaller ones into the larger ones to show them the size difference.

Place a sticker here.

Drawing one smaller

Look at each picture. Draw a smaller object beside each one.

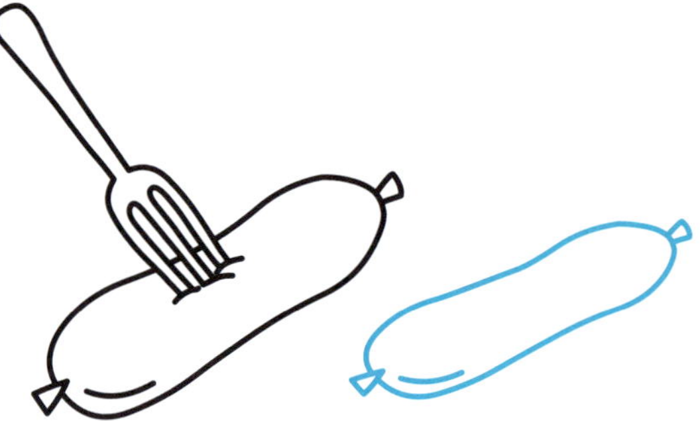

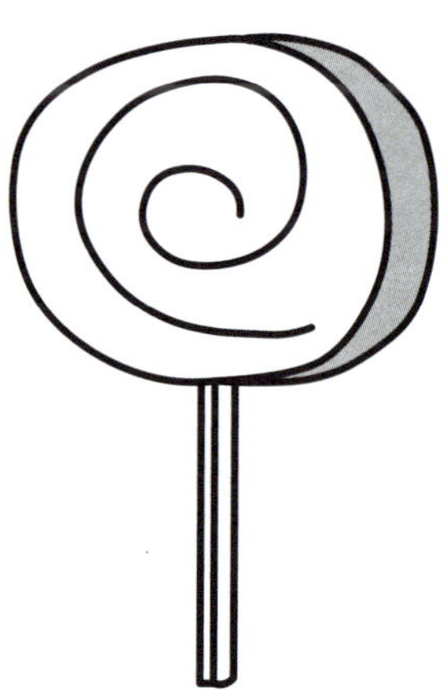

1. Ask your child to name 5 things smaller than they are, then draw them on a piece of paper.

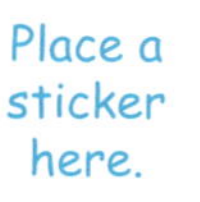

Place a sticker here.

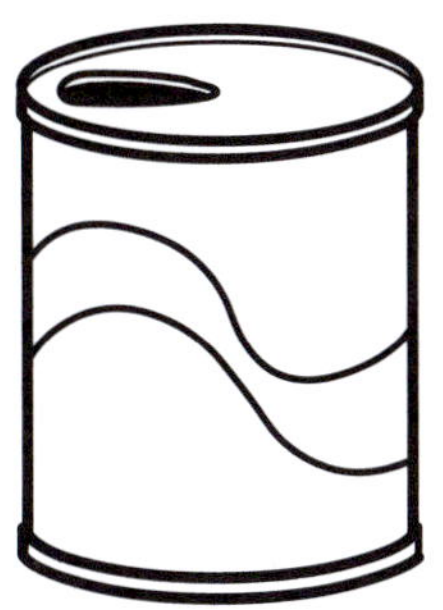

2. Ask your child which object takes up less space: your television or your refrigerator?

Place a sticker here.

Finding the lighter one

Look at each pair of pictures. Which one would feel lighter? Circle the lighter object.

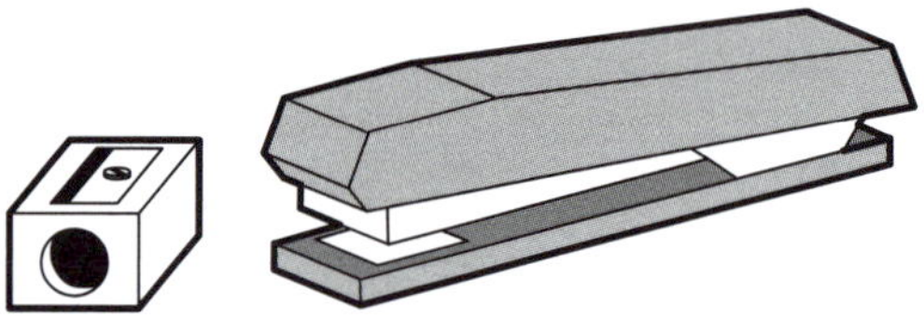

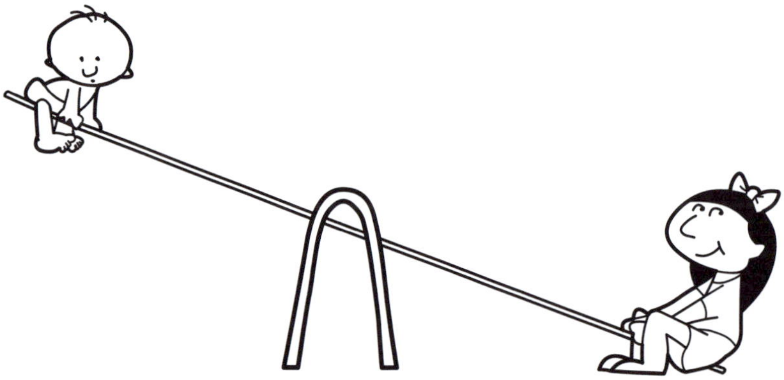

1. Find two different objects, one much lighter than the other. Have your child close their eyes and put one object in each of their hands. Ask which feels the lightest.

Place a sticker here.

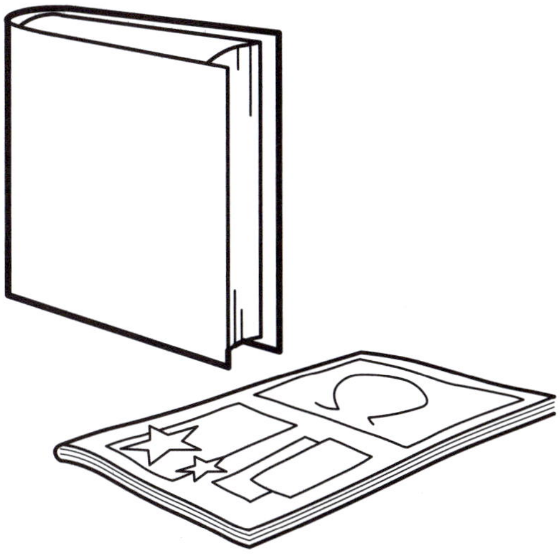

2. Try to find pairs of objects around you at home and ask your child to guess which one is lightest in each pair. Then put the objects in your child's hands to feel which is lighter.

Place a sticker here.

Finding cold things

Look at each pair of pictures. Which one shows something that is cold? Circle the cold thing in each pair. Then draw your own 'cold' and 'hot' things in the space at the bottom of each page.

Cold

1. Before your child starts this page, ask them to name 2 cold things (eg. a drink from the fridge or an ice block).

Place a sticker here.

Hot

2. Ask your child to colour all the cold things in blue and the hot things in red.

Place a sticker here.

Finding what comes last

What do you do *last* before you go to bed at night? Look at each pair of pictures. Circle the one that comes last.

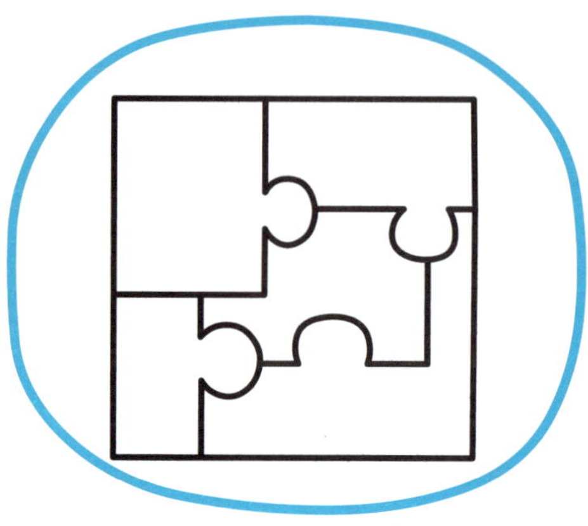

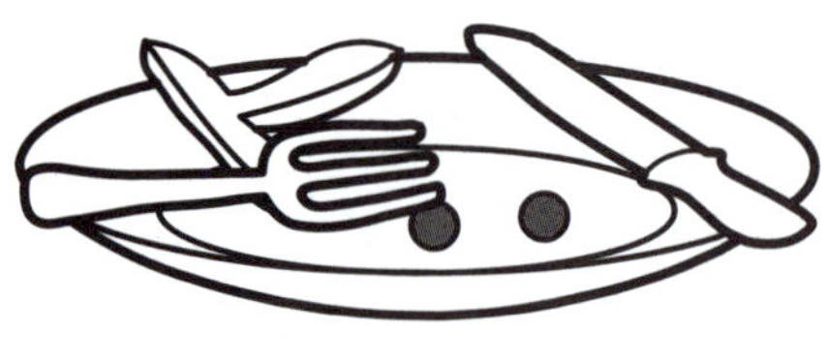

1. Ask your child whether a pet cat is a kitten before or after it is a fully-grown cat. So which one comes last?

Place a sticker here.

2. Discuss times of the day with your child. Does 1 o'clock come before or after 4 o'clock?

Place a sticker here.

Putting events into order

Look at each group of 3 pictures. Which one comes first? Which one comes next? Which one comes last? Write the number 1 under the first, 2 under the second and 3 under the third.

1. Talk to your child about your various daily activities and discuss which order you do them in. Does breakfast come before lunch? Do you eat dessert before you eat dinner?

Place a sticker here.

2. Talk to your child about the changes that occur as time passes. Will they become taller or shorter? Will they get older or younger?

Place a sticker here.

Finding cylinders

What is a cylinder? All the shapes on the first page are cylinders. Colour all the cylinders in green. Then, on the next page, draw a circle around any shape that is not a cylinder.

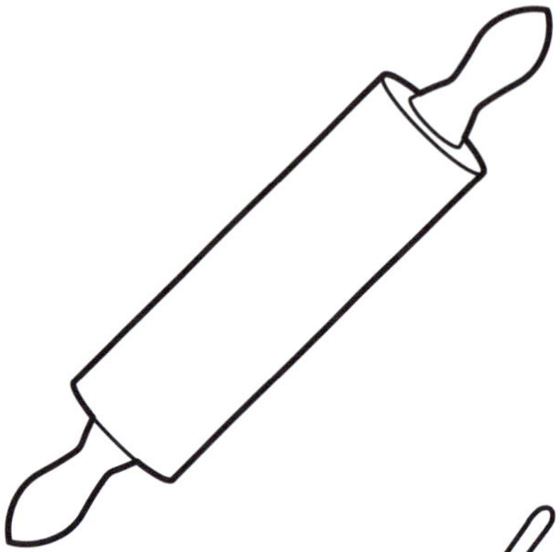

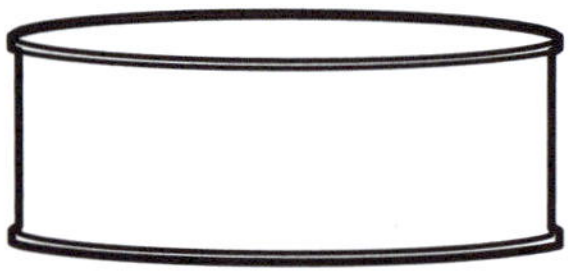

1. See how many cylinders your child can find in the kitchen or in their bedroom.

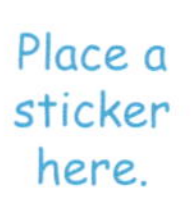

Place a sticker here.

2. Help your child make their own cylinder using rolled up cardboard or paper.

Place a sticker here.

Finding cones

What is a cone? All the shapes on the first page are cones. Colour all the cones yellow. Then, on the next page, draw a circle around any shape that is not a cone.

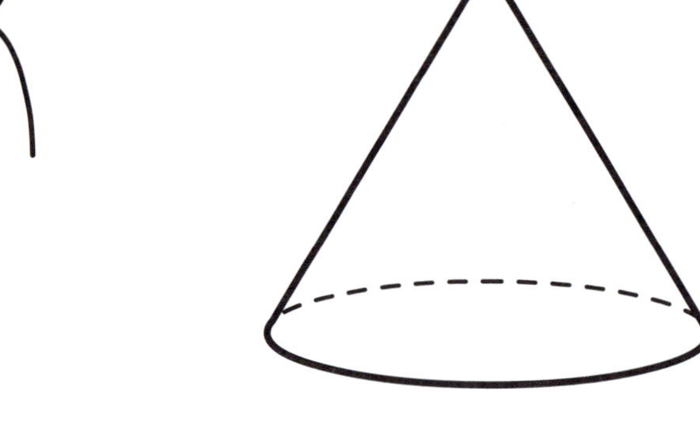

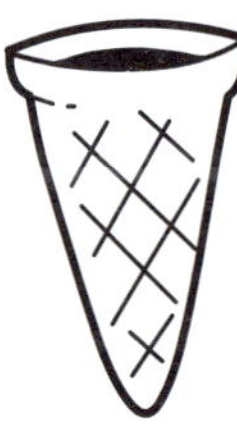

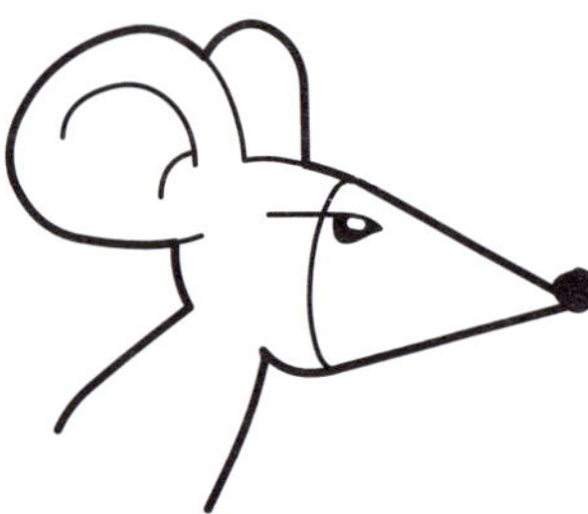

1. Help your child make a cone shape by rolling up a sheet of paper and sticking it together.

Place a sticker here.

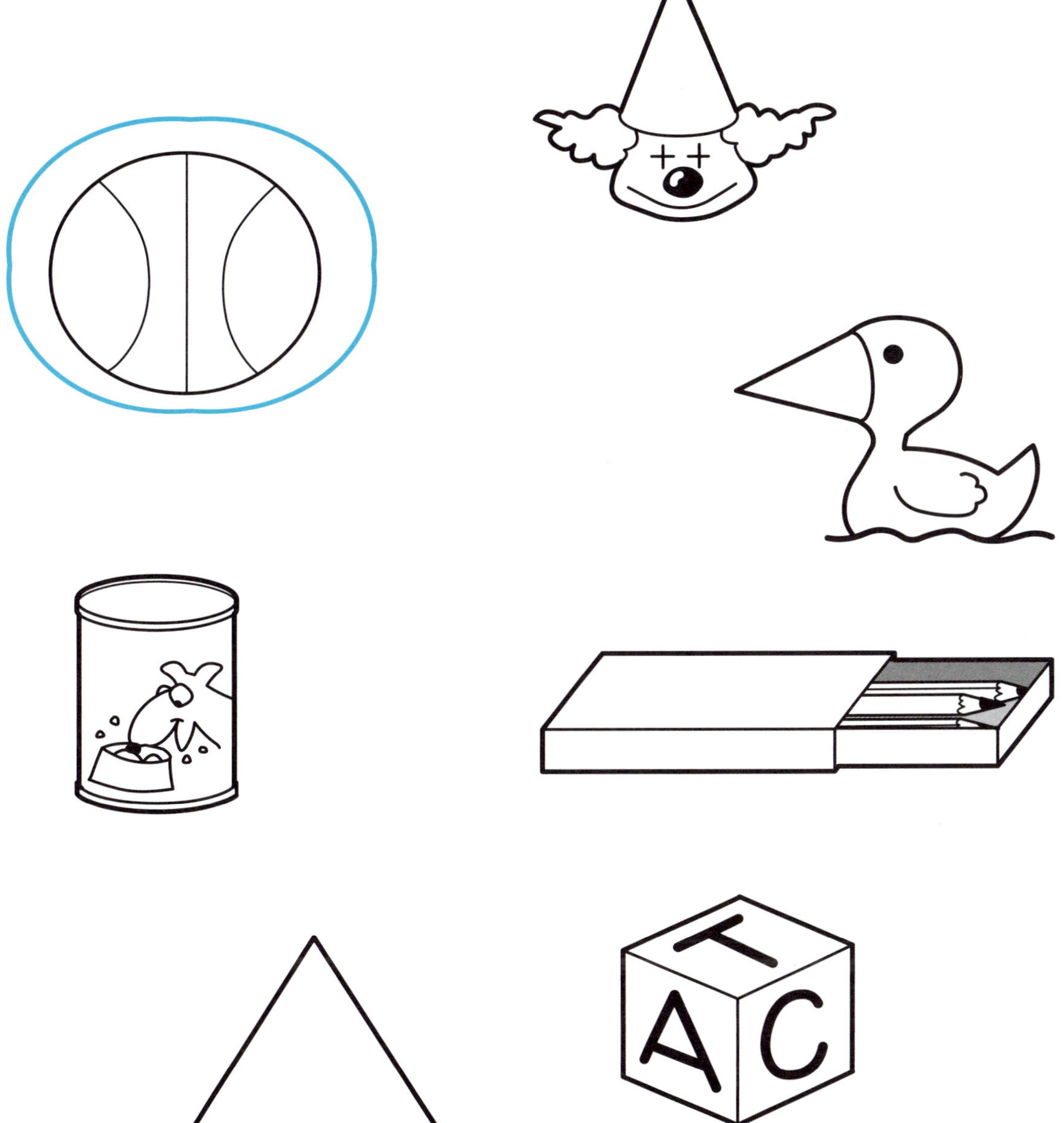

2. See if you can help your child find any cone shapes around you in the home.

Place a sticker here.

Sorting 3D shapes

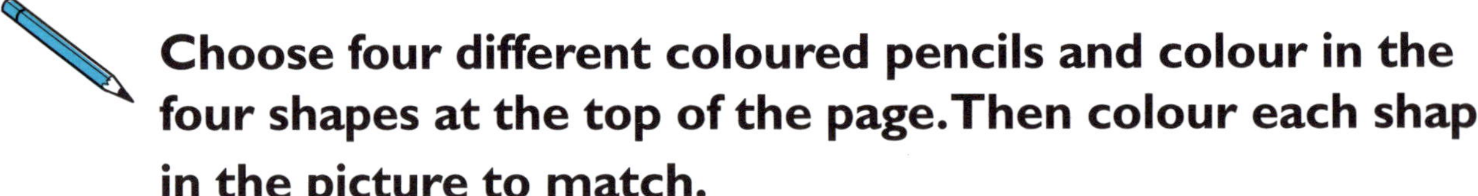

Choose four different coloured pencils and colour in the four shapes at the top of the page. Then colour each shape in the picture to match.

1. You and your child could build your own tower using objects from around you in the home. Then see if your child can name the different shapes.

Place a sticker here.

2. Explore what happens when you move cylinders, cones, boxes and balls. Do they roll or slide?

Place a sticker here.

Drawing circles

A circle has no straight sides or corners. Trace all the circles on the first page. Then draw 10 circles on the dress.

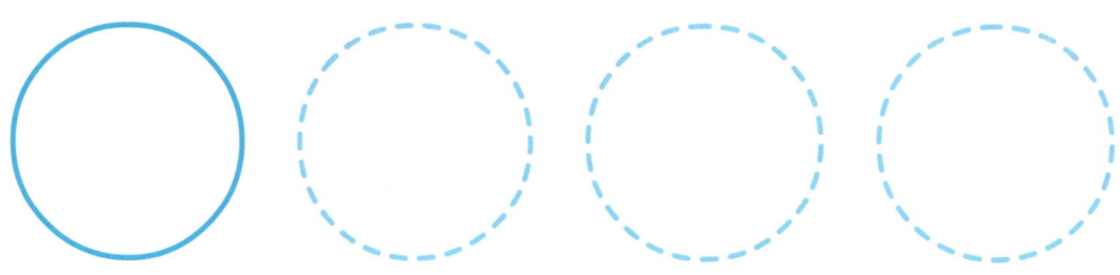

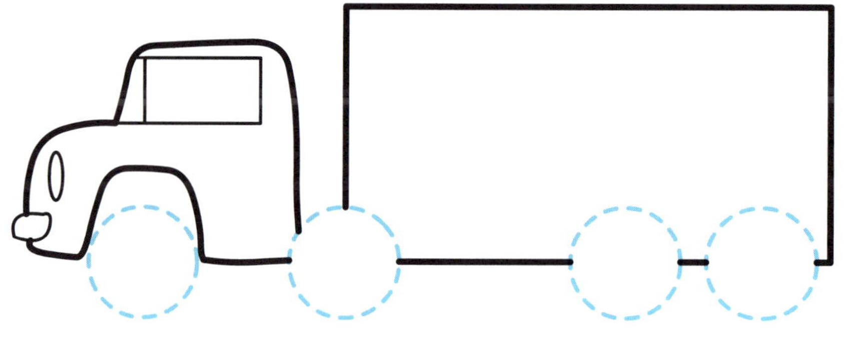

1. Ask your child to count how many circles are on this page.

Place a sticker here.

2. Help your child draw circles by tracing around a saucepan lid or small plate on a piece of paper.

Place a sticker here.

Drawing squares

A square has 4 straight sides and 4 corners. Each side is the same length. Trace over all the squares on the first page. Then draw 10 squares on the T-shirt.

1. Ask your child to count how many squares are on this page.

Place a sticker here.

2. Help your child draw a pattern of squares by tracing around a square dish or tile on paper.

Place a sticker here.

Drawing triangles

A triangle has 3 straight sides and 3 corners. The sides can be the same or different lengths. Trace over all the triangles on the first page. Then draw 10 triangles on the rug.

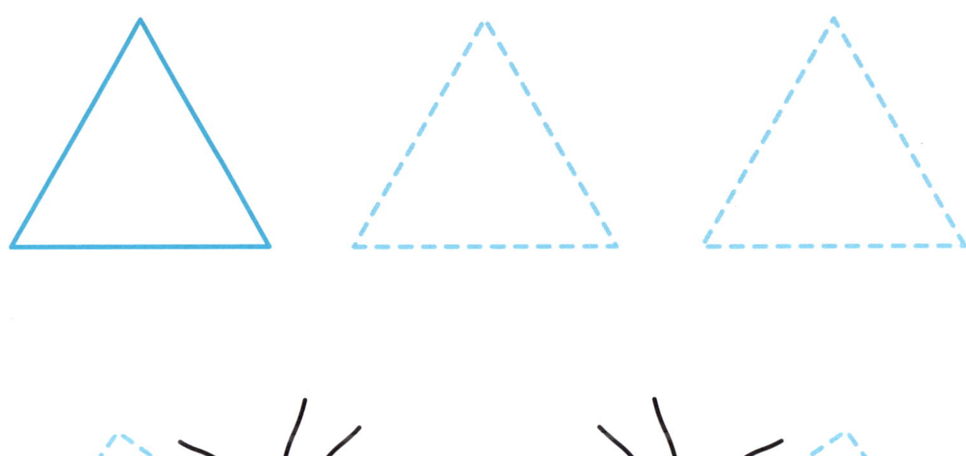

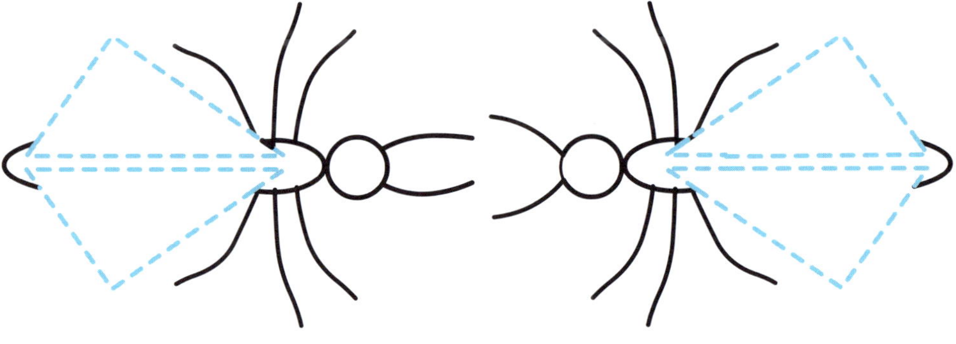

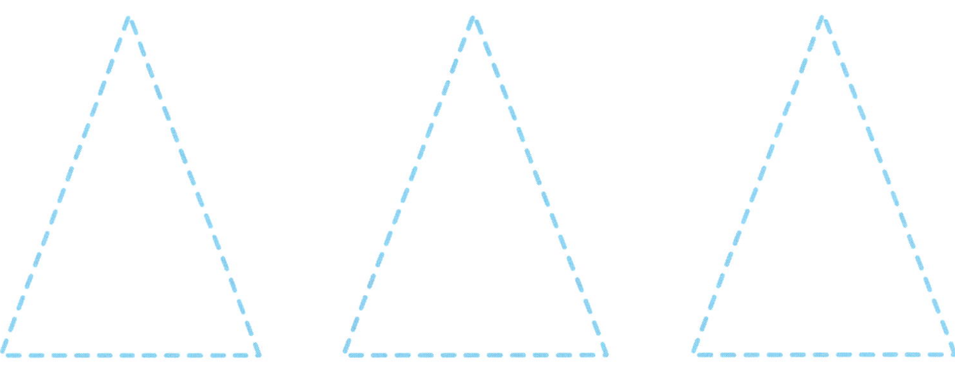

1. Ask your child to count how many triangles are on this page.

Place a sticker here.

2. Draw a series of different shapes on a sheet of paper and ask your child to colour in the triangles.

Place a sticker here.

Drawing rectangles

A rectangle has 4 straight sides and 4 corners. Two sides are long and 2 sides are short. Trace over all the rectangles on the first page. Then draw 10 rectangles on the picture frame.

1. Ask your child to count how many rectangles are on this page.

Place a sticker here.

2. You can help your child make rectangles from paper, playdough or plasticine. Where else can they see rectangles around them in the room?

Place a sticker here.

Joining the dots

What shapes can you see? Join the dots to discover each shape, then join the last number back to number **1**. What is the name of the shape you can see?

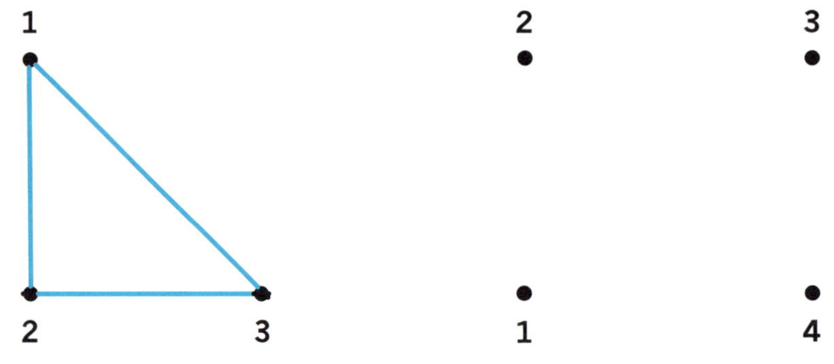

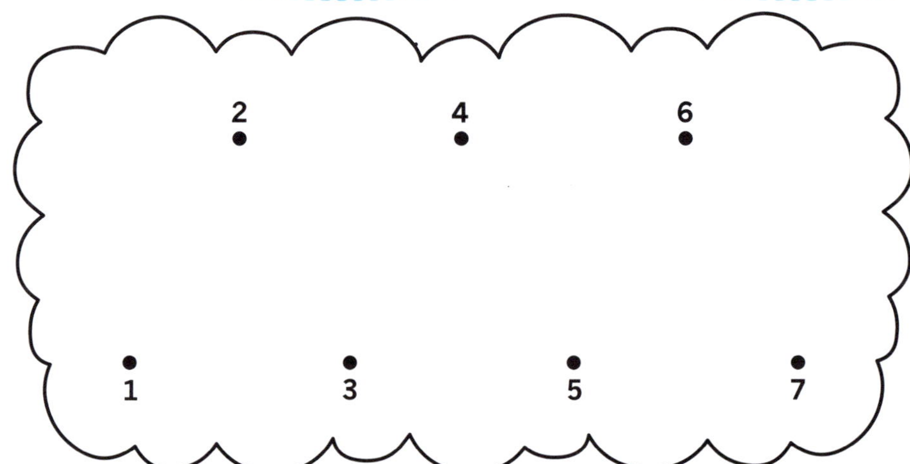

1. Ask your child to colour all the triangles pink, all the squares blue and all the rectangles orange.

Place a sticker here.

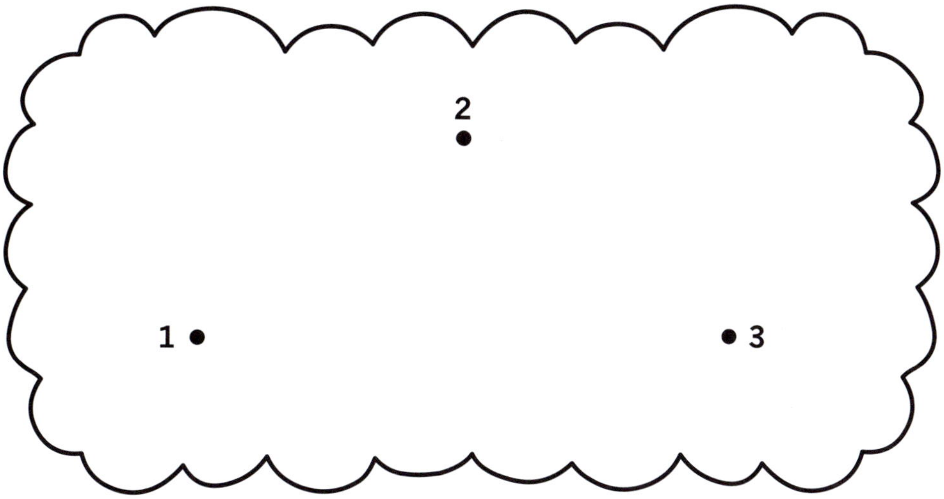

2

1 3

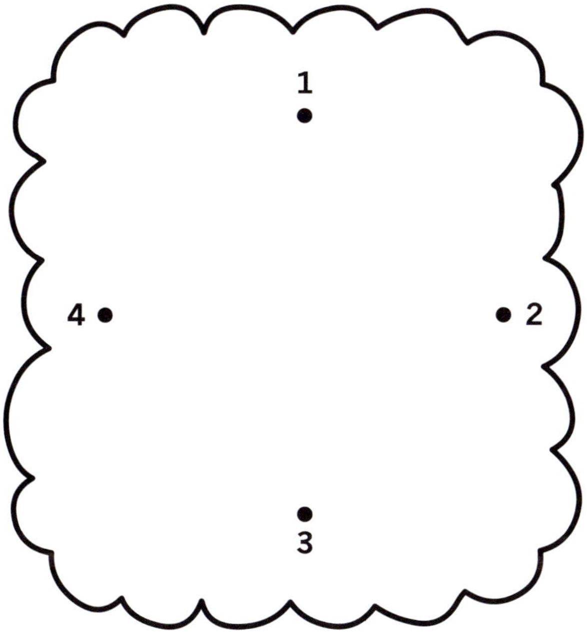

1

4 2

3

1 2

4 3

2. Help your child create their own join the dots puzzle on paper, using simple shapes.

Place a sticker here.

Drawing patterns

Trace over each pattern on the first page. Then draw the same pattern across onto the next page.

1. Ask your child to tell you about each pattern they draw. What shapes are they drawing?

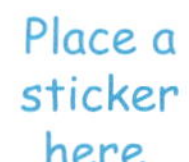

Place a sticker here.

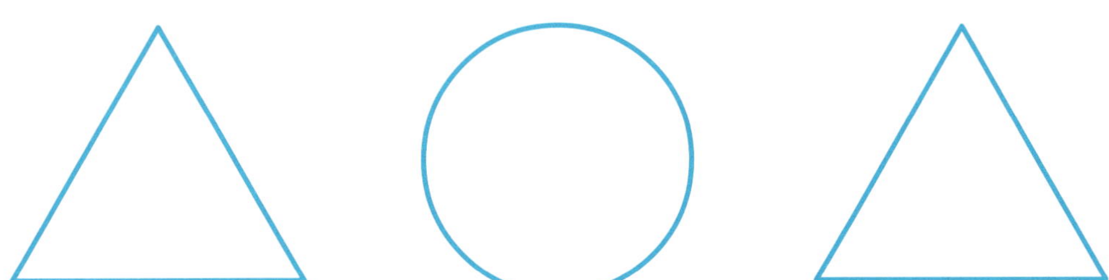

2. Ask your child to look for patterns in their room.

Place a sticker here.

Drawing a picture

You can make a picture by drawing different shapes. Trace the shapes to make a clown, and then a house with a garden.

1. Ask your child to count the circles in each picture, then count the squares.

Place a sticker here.

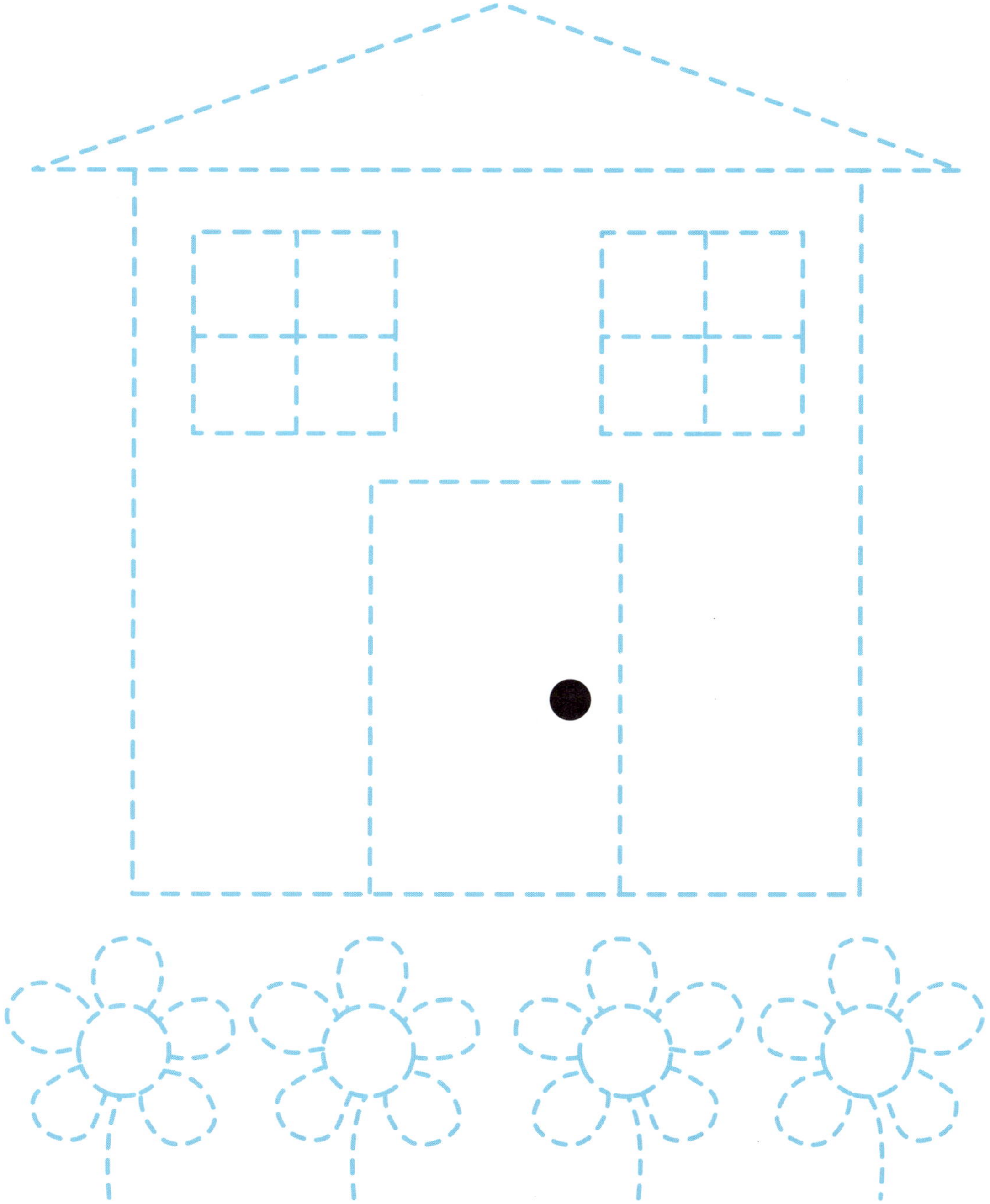

2. Help your child draw their own shape picture on paper.

Place a sticker here.

Drawing positions

'Position' tells you where you are. What can you get *on top of*, *inside* or *under*? Draw someone *on top of*, *inside* and *under* each object.

1. Next time you go out to the park or are playing outside, ask your child to stand in certain positions. For example, can they stand under the slide? On top of the roundabout? Inside the cubby house?

Place a sticker here.

2. Ask your child to draw a series of pictures following your instructions about position. For example, ask them to draw a house in the middle of the paper, a bird on top of the house, and a dog under the house.

Place a sticker here.

Finding the way (1)

In a maze, some paths lead to dead ends, and some paths lead to other paths. Find your way through each maze so the car gets to the garage and the koala gets to the gum leaves.

1. Some of the mazes may present a challenge for your child. Help them trace the correct path with their finger first before drawing it in.

Place a sticker here.

2. Ask your child to describe how they get from one room to another in your home, eg. from their bedroom to the bathroom, or from the kitchen to the front door.

Place a sticker here.

Finding the way (2)

Find your way through each maze so the dog gets to her bone and the postman gets to the letterbox.

1. Use playdough or plasticine to mould a simple maze with your child.

Place a sticker here.

2. Create a simple map on paper which shows your child a path around the park, down to the shop or around the yard. Help them follow the map as they walk.

Place a sticker here.

Well done!

You have finished the book!

**Place your last two stickers on the picture.
You can colour in the picture, too.**